SONYA'S SEEDS

by Michelle Heath Yodonis
illustrated by Sara Heath

DEDICATION

To my parents
You have helped me more in life than you'll ever know.

Mommy
Thanks especially for your help on this book.

©Michelle Heath Yodonis
All rights reserved.
Published in the United States of America
December 2017

ISBN-13: 978-1979934923
ISBN-10: 1979934924

Sonya, a little girl
I used to know,
loved to sit in the wind
and watch
the flowers
blow.

Till one day
she became very curious
and thought their
existence was mysterious.

While the wind blew the delicate flowers,
Sonya sat in wonder for hours and hours.
She did not understand how the flowers grew.
That they were bright and beautiful was all she knew.

Her curiosity became too much to bear, so, she returned home to see if the answer was there.

While running into the kitchen she asked with excitement, "Mommy, where do the flowers come from that bring me such delightment?"

Mrs. Borroel turned to Sonya with a puzzled look on her face.

"Where do the flowers come from?" Sonya repeated with grace.

Her mother responded,
"They come from the ground."
Sonya smiled for she knew
her answer was found.
But soon there was another
question to be asked,
so, she returned to her
mother to fulfill the task.

Sonya questioned, "How are the flowers grown?"
Her mother replied, "From seeds that are sown."

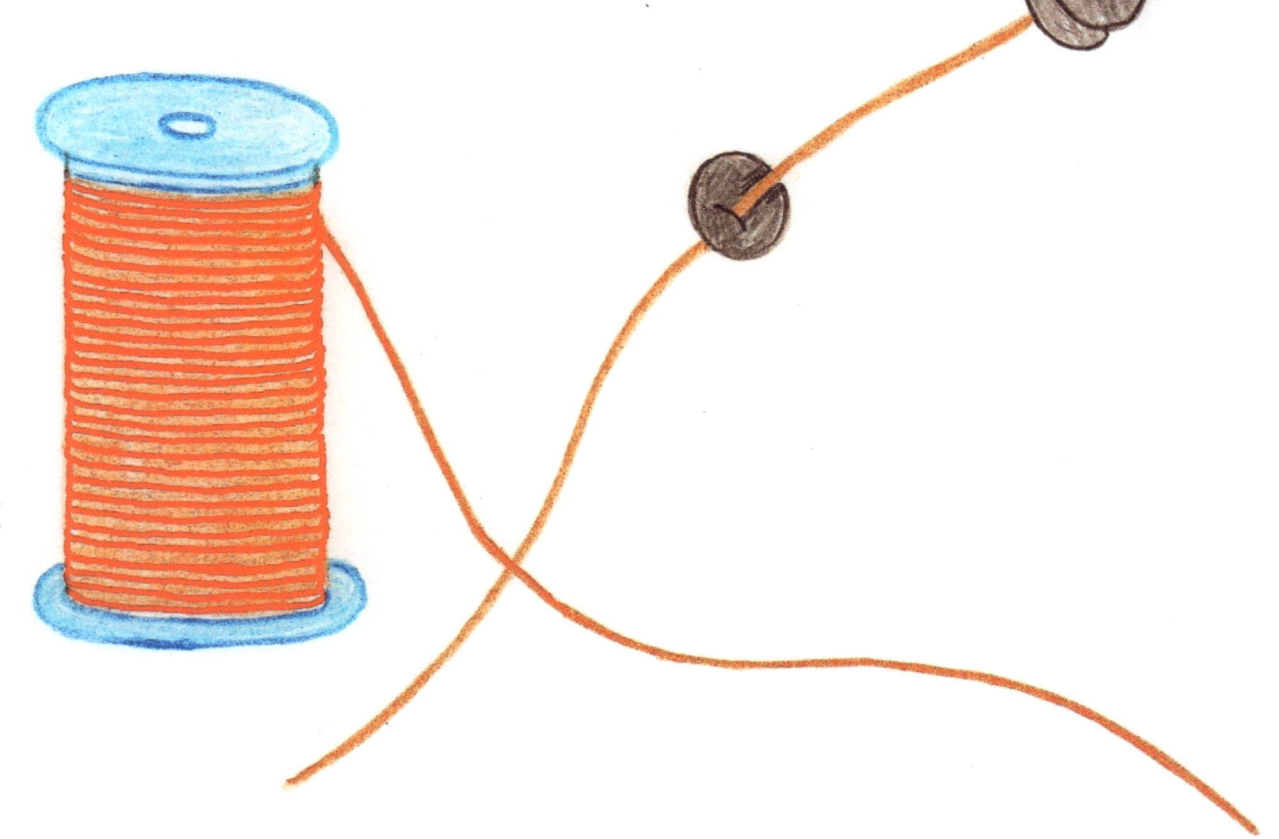

Sonya's picture of sowing seeds was not very clear. So, her mother said, "Let's go outside, dear."

Sonya followed her mom to a garden near a tree. "This is where we will be sowing our seeds, see! Sow means to plant," Mrs. Borroel explained. She showed her daughter the seeds she claimed.

Sonya's mother dug
a row of holes
with her finger.
She said,
"This is where
the seeds will linger."

Mrs. Borroel told Sonya to place a seed in each hole. Then she covered the seed with soil to protect each new soul.

Over the mound of soil, water will pour.

Mrs. Borroel said, "Soon the flowers will soar."

A few weeks passed since Sonya had sown her seeds, but she knew soon they would blow in the breeze.

New things were discovered with each passing day. Sonya knew a lot and had much to say. Green leaves sprouted while the weeks passed. For Sonya this process was not very fast.

MAY

Sunday	Monday	Tuesday	Wednesday	Thursday	Friday	Saturday
			1	~~2~~	~~3~~	~~4~~
~~5~~	~~6~~	~~7~~	~~8~~	~~9~~	~~10~~	~~11~~
~~12~~	~~13~~	~~14~~	~~15~~	~~16~~	~~17~~	~~18~~
19	20	21	22	23	24	25
26	27	28	29	30	31	

Finally, on one of
Sonya's daily checks,
she saw the colors
of the flowers,
just little specks.

She knew that soon,
in the wind,
the flowers would blow
each and every
single one in that row.

The next day when Sonya came out to play, each flower was different in its own special way.

While learning flowers
need soil, water, and sun,
Sonya had enjoyed
each day and
had tons of fun.

Thanks to her mom,
they'd planted seeds
in the ground,
and now Sonya's answer
to her question was found.

About the Author

Michelle Heath Yodonis was born in a small town in Florida. When she wrote this story she was sixteen years old.

After graduating high school, Michelle went on to college and graduated with a bachelor's in Business Management and a master's in Public Administration. She enjoys a career in public service.

She is married and has two wonderful children.